Sound

Exploring the Energy You Hear

DEVELOPED IN COOPERATION
WITH
ARIZONA MUSEUM OF SCIENCE AND TECHNOLOGY
PHOENIX, ARIZONA

Published by Scholastic Inc. Printed in the U.S.A.
ISBN 0-590-27642-5
2 3 4 5 6 7 8 9 10 09 01 00 99 98 97 96 95

THE PHYSICAL WORLD IS GOVERNED BY PROPERTIES AND INTERACTIONS OF MATTER AND ENERGY.

Sound

Sound energy is produced by vibrating objects and travels through matter in waves.

Read-Aloud

Sound

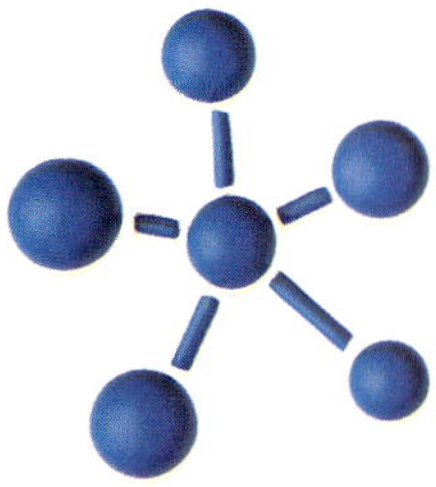

Vibrations create sound waves that can travel through, bounce off, and be absorbed by matter.

Literature

Literature

The perception of sound, including pitch and volume, depends on the properties of the sound waves.

People control sound in many ways.

What Did We Learn?

What Makes Sound?

Whoosh! Bang! Clatter. Mmmm... What could have made those sounds? Even sounds that are very different are all alike in one important way. All sounds are made by a certain kind of motion. What kind of motion makes sound?

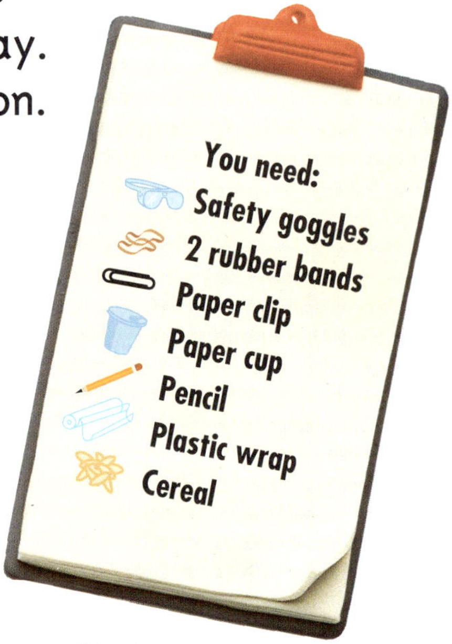

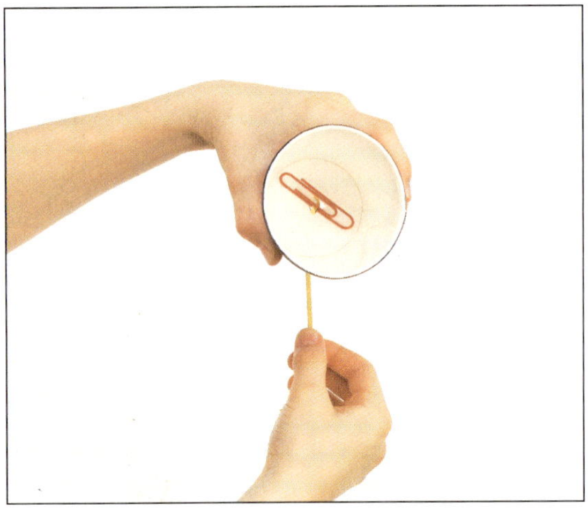

Be a sound detective.

❶ Cut one rubber band and tie it to the paper clip. Then run the band through a hole in the bottom of the cup.

❷ Hold the cup while your partner holds and gently plucks the band. What do you hear? What do your fingers feel?

❸ Stretch plastic over the top of the cup and hold it there with a rubber band. Then sprinkle cereal on the plastic. What do you see when your partner plucks the band? How can you tell when sound is made?

Movements up and down or back and forth are called vibrations. How could you tell that the rubber band was vibrating?

Inside your throat you have vocal cords. They're like two leathery flaps. Normally they have a space between them so you can breathe. But when you use your voice, your vocal cords move close together. Then air coming up from your lungs moves between them.

What do you think the air does to your vocal cords? Place your hand near the bottom of your throat and say "Ah."

What Causes Vibrations?

You know that vibrations cause sound, but what makes things vibrate? In the last lesson, you made a rubber band vibrate with your fingers. Now think about a doorbell. What do you think causes the vibrations that make a doorbell ring?

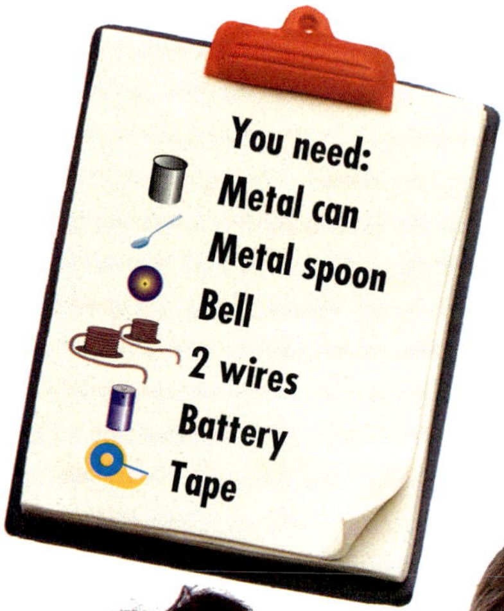

Ring your bell.

1. Use a metal can and spoon to make a ringing sound. What do you feel the can doing to produce the sound? What did you do to the can to get that sound?

2. Use the rest of the materials to complete an electric circuit and make the doorbell ring. How can you break the circuit to make it stop ringing.

3. Make a sign that gives directions for using the doorbell.

If you wanted to make a ringing sound over and over again, which would be easier to use — the can and spoon or the doorbell? Why? What made the doorbell vibrate if you didn't have to hit it or shake it yourself?

It takes energy to make something vibrate. That energy can come from you or from something else. Batteries have chemicals in them that make electrical energy.

Think about the sounds a motorcycle makes. Now think about the sounds a waterfall makes. Where does the energy come from to make those sounds?

How a Battery Works

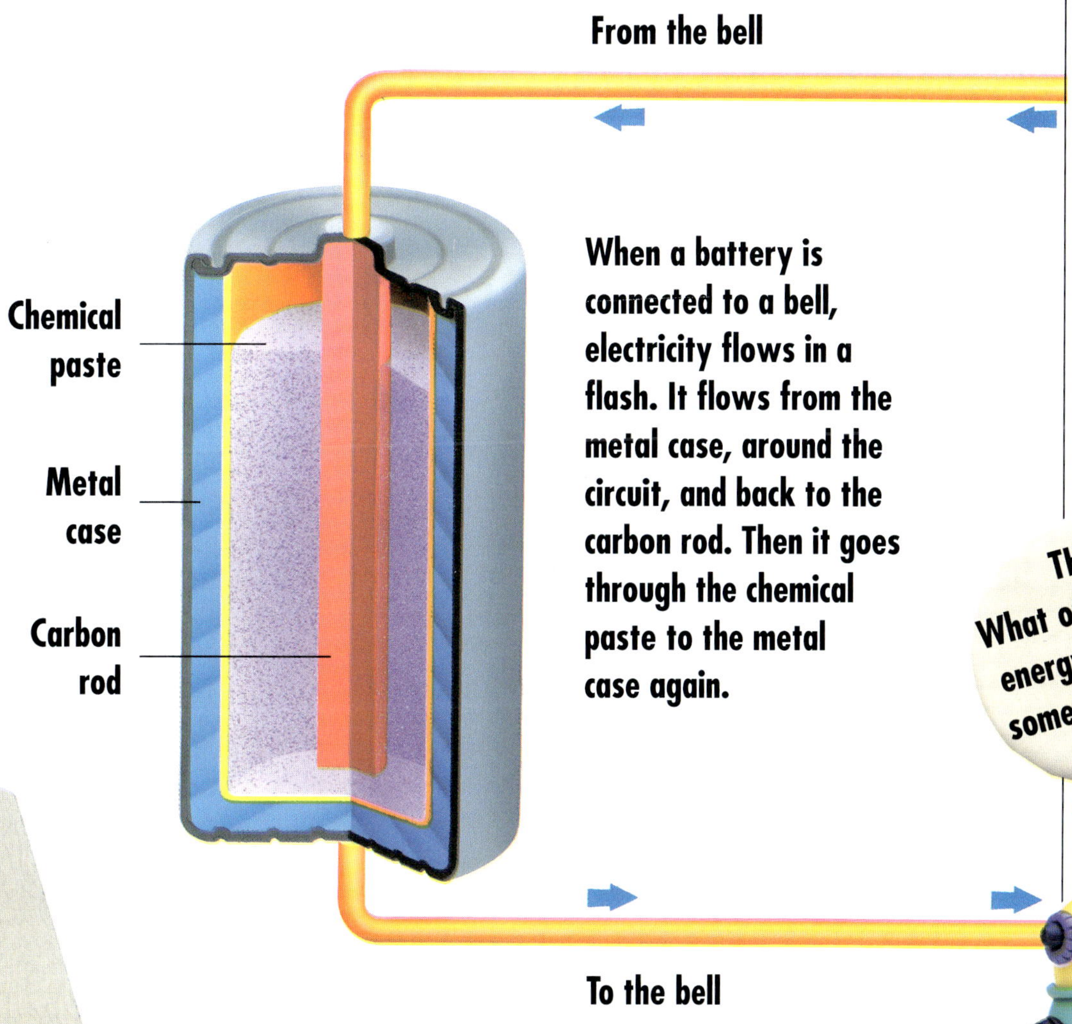

When a battery is connected to a bell, electricity flows in a flash. It flows from the metal case, around the circuit, and back to the carbon rod. Then it goes through the chemical paste to the metal case again.

How Do Vibrations Reach Your Ears?

A doorbell rings. The people who are at home hear it. What's between the doorbell and the people who hear it? How do vibrations get to your ears where you hear them as sound?

Make a vibration model.

1. Stand in a straight line with each person behind the next. Everyone should be facing the same way.

2. The person at the end of the line is a doorbell. That person taps the back of the next person and so on.

3. When the person at the very front of the line feels a tap, he or she taps a pencil on an upside-down paper cup. The pencil tapping the cup is a model of the sound reaching your ear.

4. Send a softer vibration by tapping very lightly. What should happen to the sound?

In your model, what is between the doorbell and your ear? If you rang a bell in the classroom right now, how would all of you be able to hear it?

Use your model to explain how the vibrations from this crashing cymbal reach your ears.

When something vibrates – like a cymbal – it pushes the air around it. That air pushes more air and that air pushes more air until the air inside your ear gets pushed, too. That motion between the cymbal and your ear is a sound wave.

THINK!
What do you think happens to the energy it takes to start a vibration?

How Do Sound Waves Travel?

The sound of a bell ringing near you easily reaches your ears. But suppose you heard an alarm from a parked car during the night. At school the next day, people who live blocks away say they heard the alarm too. How do you think many people hear the same sound?

In which directions are the sound waves from this noisy alarm traveling?

You need:

Paper clip

Track the sound.

1. Listeners stand facing outward in a circle around a sound maker.

2. The sound maker drops a paper clip. All listeners who hear the sound raise their hands. Then they take two giant steps forward.

3. Repeat step 2 until no one can hear the sound. What happens to the sound as the listeners move farther and farther away? Draw what you think happens to the sound waves.

Sound waves travel away from the source of the sound in straight lines. They travel in many directions at the same time. How can you tell which directions sound waves travel in?

THINK!
If a bell rings to tell you when recess is over, why can't the sound be heard miles and miles away?

What Can Sound Waves Travel Through?

Sound waves travel through air when you listen to a radio from across a room. If the music's loud enough, you can feel the beat vibrating in the floor. If it's really loud, you can hear it from another room. What are the radio's sound waves traveling through?

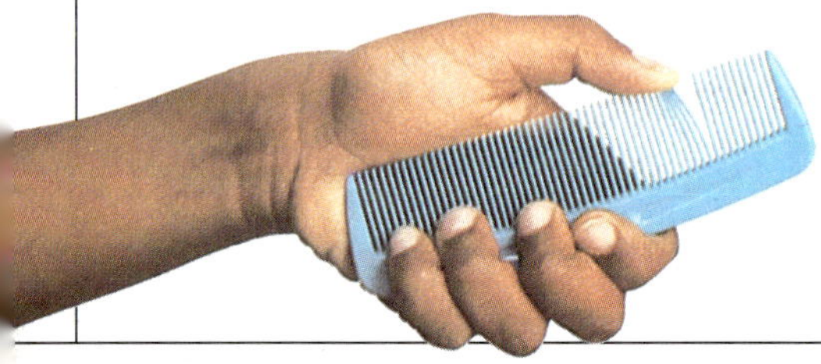

Make sound waves travel.

❶ Strum the teeth of a comb with your thumb. Listen to the sound. What do the sound waves travel through to get to your ear?

❷ Strum the comb against the edge of a desk. Take turns with your partner listening with one ear on the desk. How does the comb sound? What do the sound waves travel through now?

❸ Strum the comb in the container of water. Take turns with your partner listening with one ear against the container. How did the sound change? What did the sound waves travel through here?

Sound waves must travel through matter to reach you. A rocket taking off from Earth makes a loud noise because the sound of the rocket moves through air to get to you. But there's no air on the moon. If you were standing on the moon near a rocket taking off, do you think you could hear it? Why or why not?

Like you, fish can hear sounds underwater, too. Some fish, such as catfish, send sounds to one another. How do you think fish called grunts and croakers got their names?

How Do Your Ears Catch Sound Waves?

Sound waves are traveling around you in all directions. How do your ears help catch these sound waves? Can you tell where each sound comes from without looking? How?

Find that sound.

1. Have your partner hold one ear closed and sit in a chair with his or her eyes closed.
2. Clap your hands around your partner's head — to the left, the right, above, and below. Ask your partner to keep his or her head still and point to where the sounds are coming from. Keep a record of the answers.
3. Repeat step 2 while your partner listens with both ears. How accurate is your partner now?
4. Repeat step 2 and 3 while your partner cups his or her ears. What difference does your partner notice?

How does having one ear on either side of your head help you figure out where sounds are coming from?

How Sound Waves Get to Your Brain

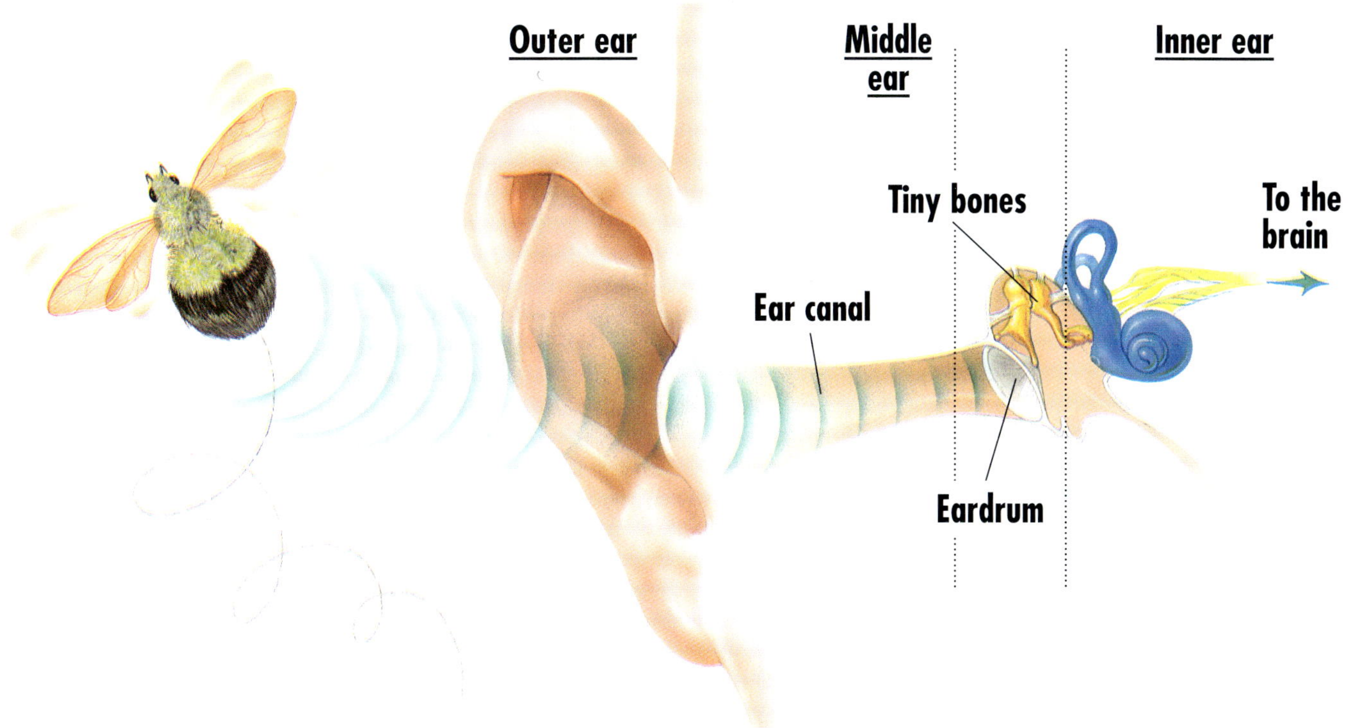

Sound waves travel into your ear canal, and make your eardrum vibrate. Your eardrum makes the tiny bones in your middle ear vibrate. Then the bones send the vibration to your inner ear. Your inner ear turns the vibration into a signal your brain can understand.

A hearing aid that's worn behind the ear can help a person whose middle ear doesn't work.

Hearing aids can help some people who have trouble hearing. A certain kind of hearing aid turns sound waves into vibrations the same way your eardrum does by hitting the bone behind the ear. Then the vibrations travel through the bone to the inner ear.

How Do Other Animals' Ears Work?

Think of different animal ears and how they look. Which animals have big ears? Which animals don't seem to have any ears at all?

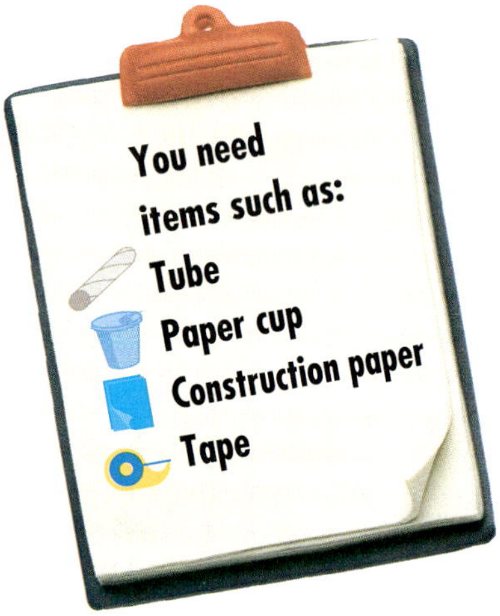

Lend me your ears.

❶ Make outer ears that will help you hear better.

❷ Work in groups. Make sounds to test your new ears. How do they work best for catching sound waves?

You might think that a cricket has no ears — unless you know where to look. A cricket hears with a vibrating membrane — a thin piece of skin — on each front leg.

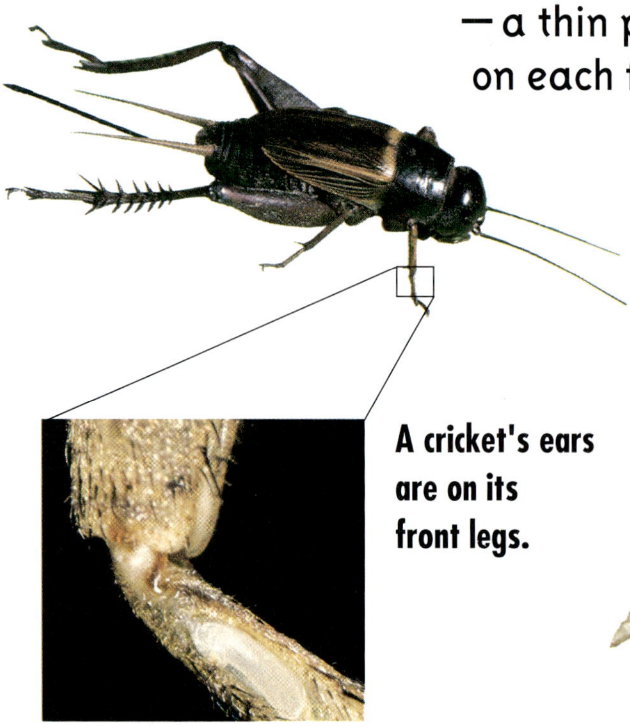

A cricket's ears are on its front legs.

A fox can hear sounds coming from under the ground.

Some animals have ears that can do things human ears can't. A fox's large ears turn to hear sounds coming from any direction. Which animal do you think has the best hearing: a cricket, a snake, or a fox? What do you think makes it so?

Snakes don't have ears you can see. But a snake can hear by putting its head on the ground. Sound waves traveling through the ground make the bones in the snake's head vibrate. The bones send the signal to the snake's inner ear.

A snake has no outer ears but can hear!

What Makes the Sounds You Hear Soft or Loud?

Some animals can hear very soft sounds. What kind of vibration do you think makes a soft sound? What kind do you think makes a loud sound? The loudness or softness of a sound is its volume.

Try clapping loudly. Now try it with your hands close together. What do you think makes a sound's volume softer or louder?

Which has more energy – the sound a waterfall makes or the sound of water dripping?

You need:

- Safety goggles
- Rubber band
- Cup covered with plastic wrap
- Cereal
- Construction paper
- Tape
- Scissors

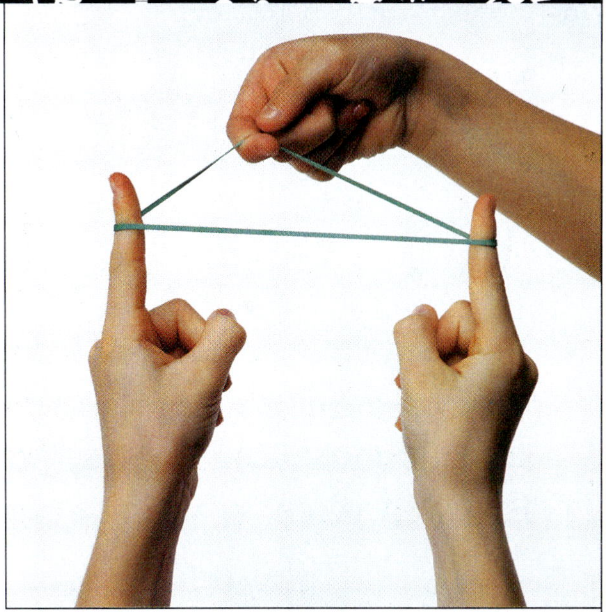

Turn it up.

1. Have your partner stretch the rubber band between two fingers.

2. Gently pluck the rubber band. How big was the vibration? How loud was the sound you heard? How can you make the sound louder?

3. Sprinkle some cereal over the plastic wrap on your cup.

4. Make a paper megaphone. Point it down at the cereal and talk. What happens? What happens when you talk louder and louder? Record what you see.

What kind of vibration did it take to make the rubber band sound loud? Did soft or loud sounds make the cereal vibrate more? Which are more powerful — loud or soft sound waves? Which do you think travel farther?

What Makes the Sounds You Hear High or Low?

Sounds can be soft or loud. But sounds can also be high or low. How high and low can you make your voice? The highness or lowness of sound is called pitch. What makes different pitches?

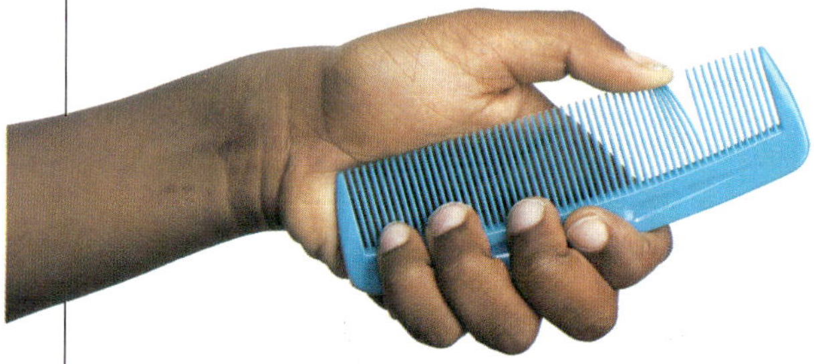

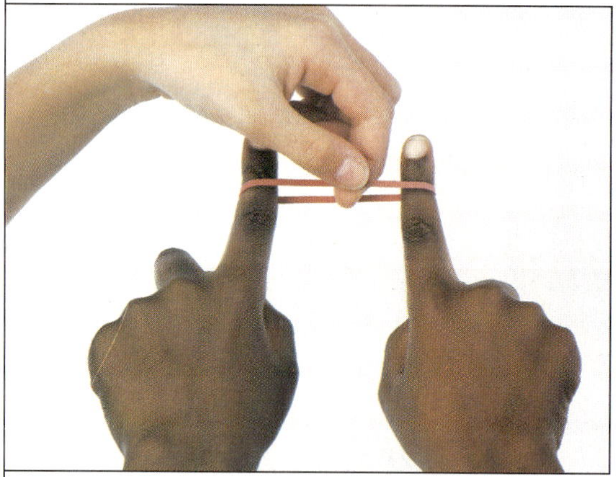

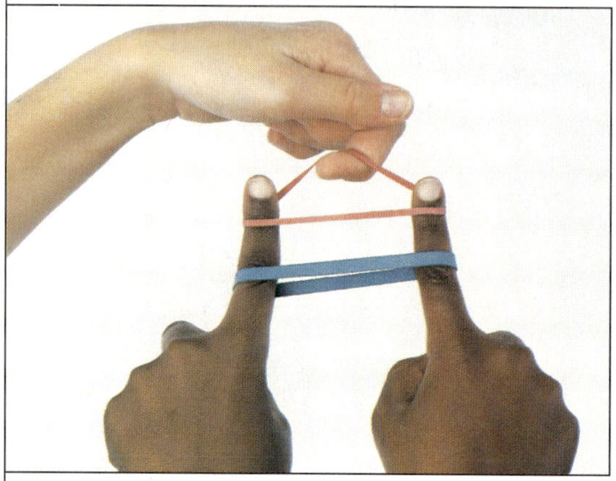

Play it high and low.

❶ Strum the teeth of a comb with your thumbnail. Do it faster. Do it slower. How does the sound change?

❷ Have your partner stretch a rubber band around two fingers. Pluck it. How does it sound when the rubber band is tightened? When it's loosened?

❸ Next, shorten the vibrating part of the rubber band by pinching it in the middle. How does that change the sound when you pluck it?

❹ Now have your partner stretch a second rubber band around the same two fingers. Which sounds higher — the thin one or the thick one? Make a table of your results.

Which do you think vibrates faster, a rubber band that's thick or thin? One that's long or short? One that's tight or loose?

Cello strings

Which do you think vibrate faster—violin or cello strings?

The number of vibrations an object makes in one second is called its <u>frequency</u>. A high frequency — a lot of vibrations per second — makes a high-pitched sound. A low frequency makes a low-pitched sound. Which do you think makes more sound waves in one second — a high frequency or a low frequency?

What Sound Frequencies Can Animals Hear?

You can hear a robin's high-pitched tweet, a dog's low-pitched growl. But how high and low are the sound frequencies robins and dogs hear? Do all animals hear the same frequencies? Use the table on the next page to see.

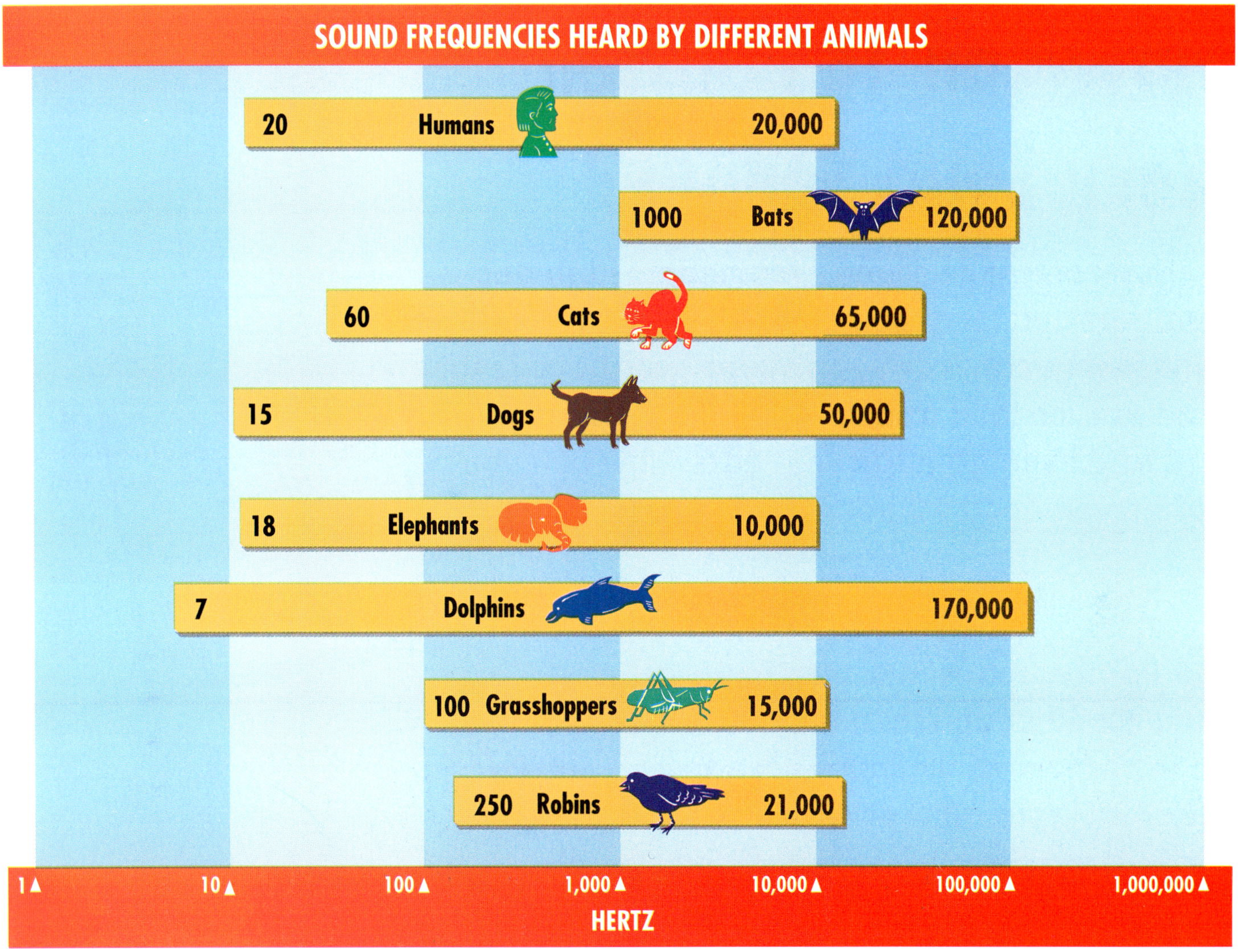

Which animals hear lower sound frequencies than humans can? Which animals hear higher ones?

Scientists measure sound frequency in hertz (Hz). The number of hertz is the number of vibrations made in one second. When you make a 500 Hz sound with your voice, your vocal cords vibrate 500 times in one second. When you hear a 500 Hz sound, how many times do you think sound waves hit your eardrum in one second?

How Can You Soften Very Loud Sounds?

Think of all the sounds you hear — high and low, soft and loud. Some loud sounds are useful. Which ones can warn you of danger? But sounds that are too loud and last too long can bother you and even damage your ears. How can you soften very loud sounds?

You need materials such as:
- Beans
- 2 cups
- Tape
- Small box
- Cotton balls
- Newspapers
- Foil

Turn it down.

1. Put the beans in a cup. Tape the cups together and then tape the cups to the bottom of the box. Shake the box and listen. Now cover the box and listen.

2. Next, try putting other material in the box. Which material soaks up the most sound? Compare your results with your classmates'.

The Decibel Levels of Different Sounds

Rustling leaves
10 db

A whisper
20 db

Normal conversation
60 db

Vacuum cleaner
70 db

Noisy cafeteria
80 db

Moving train
90 db

Rock concert
110 db

Jet taking off nearby
140 db

Volume is measured in decibels (db). The number of decibels tells how powerful a sound is. But the difference between decibels is greater than you might think. A sound going from 10 to 20 decibels gets 10 times as powerful. One going from 10 to 30 decibels gets 100 times as powerful.

Listening to sounds over 90 decibels for more than 8 hours at a time can damage a person's hearing. Sounds of 140 decibels are not safe to listen to for even a second. Which sounds listed in the table could be harmful? Loud and constant noise is called noise pollution.

THINK!
How can you soften the sound of an alarm clock ringing?

How Can You Send Sound Waves Where You Want Them to Go?

You know that you can send a loud shout farther than a soft whisper. But suppose you shout "Wait up!" to some friends who are walking ahead of you. You yell as loud as you can and they still don't hear you. How could you help your voice reach them?

You need:
- Construction paper
- Tape
- Scissors

Send your sound.

❶ Work in groups to make the megaphones. Each group's megaphone should have a different-sized opening at the large end.

❷ The speaker stands in the middle of a room. Listeners spread out in a circle, facing the speaker.

❸ The speaker whispers hello to the listeners. All those who hear the whisper raise their hands.

❹ Then the speaker whispers into each megaphone. The megaphone should point in the same direction each time. How do the whispers change? How many listeners hear the whisper each time?

Which instruments can direct the sounds they make toward or away from the audience?

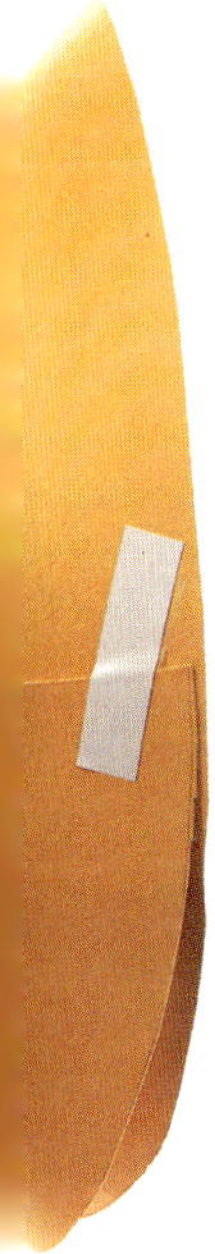

What did the megaphones do to the sound? What did they do to the sound waves? What do you think would happen if you turned the megaphone around and spoke into the large end?

Many musical instruments have ends that are shaped like megaphones. How do you think this shape could change what the audience hears?

How Can You Use Sound Waves for More Than Hearing?

You controlled sound when you made it softer or sent it in the direction you wanted it to go. But scientists have found ways to do even more with sound. Look at the pictures to see some amazing ways people can use sound.

Have you ever wondered what you look like inside? Sonogram machines can take pictures of your insides by using echoes. Sound waves go through your skin and when they hit something inside, they bounce back. It doesn't hurt a bit.

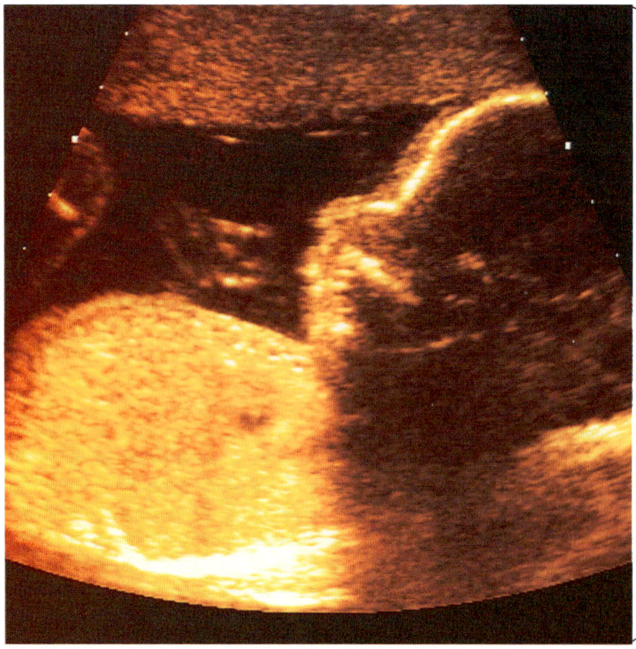

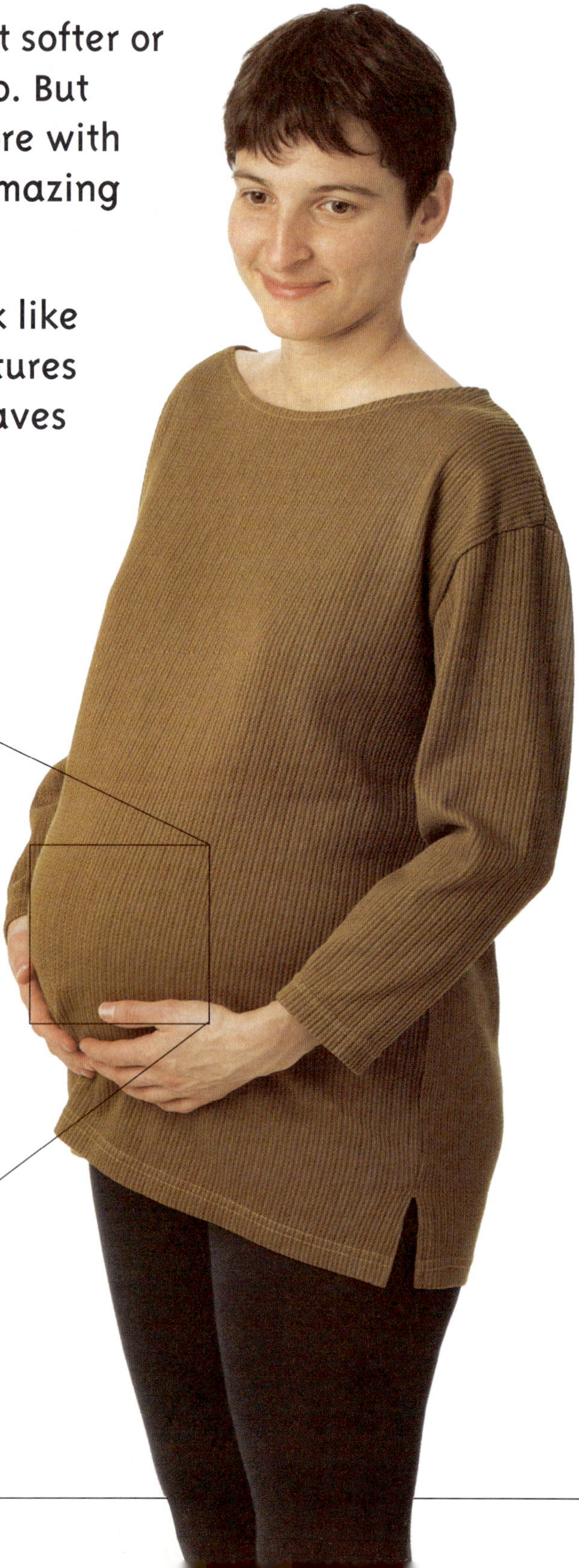

Doctors often use sonograms to see unborn babies inside their mothers' bodies.

The pattern of sound waves that your vocal cords make is like your thumbprint — no two are alike in the world.

There's a machine that can take the sound waves from your voice and make a picture of them. This machine is called a sound spectrograph. How do you think a machine like this could someday take the place of door keys?

Sound can also be used for cleaning. The delicate parts inside a watch would break if you scrubbed them. But quick vibrations made by a very high-pitched sound can shake off tiny pieces of dirt and dust without breaking the watch.

What does sound have in common with a scrub brush?

THINK!
How could sounds be helpful in mapping the bottom of the ocean?

How Can You Combine Objects to Make Music?

Sounds can do many amazing things. They can even change the way you feel. Which sounds make you happy? Which make you sad? Think of all the objects you can combine to make different kinds of sounds. How can you put these sounds together to make music?

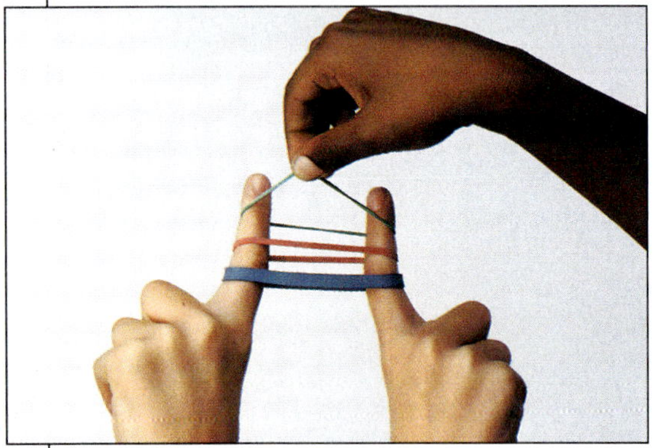

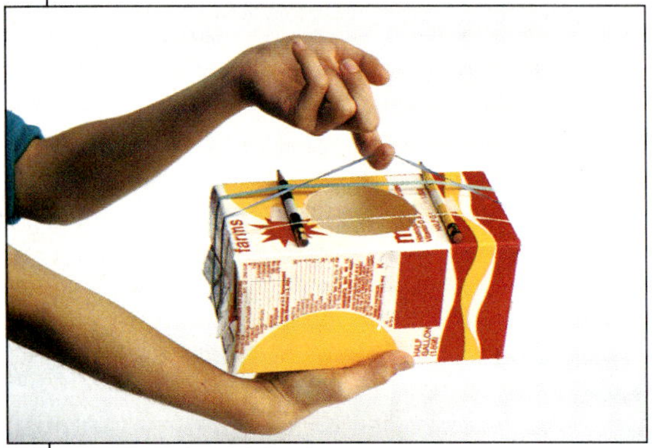

Change your tune.

❶ Pluck three rubber bands and listen to the sounds.

❷ Stretch the rubber bands evenly around two closed boxes. Tape two pencils between the rubber bands and the top of each box. How does the box change the plucking sounds?

❸ Now make a <u>sound box</u> by putting a hole in the top of one box. Compare the sounds each box makes.

❹ Work in a group to make other instruments. Everyone's instrument should be different. What sounds does your instrument make? How do your instruments sound as a group?

Musical instruments can be divided into big groups. There are string instruments, such as a violin; wind instruments, such as a flute; and percussion instruments, such as a bongo drum. How would you classify your instrument?

Does your instrument have a sound box like the one you made for the rubber bands? What do you think vibrates when you use a sound box? What happens to the sound waves? Do you think the box bounces them back or soaks them up? Which parts of your instrument vibrate? How can you change the pitch? How can you change the volume?

GLOSSARY

Decibel (db): Decibels measure the volume of a sound vibration. A sound of 0 decibels is the weakest sound the human ear can hear.

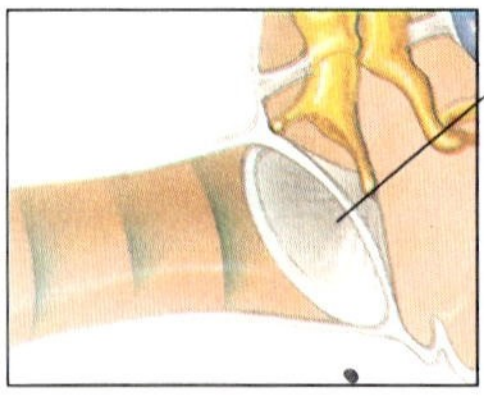

Eardrum: The eardrum is a membrane at the end of the ear canal. It's as thin as tissue paper and vibrates when sound waves enter the ear.

Frequency: Frequency tells how fast something is vibrating. The faster something vibrates, the higher its frequency.

Hertz (Hz): Sound frequencies are measured in hertz. The number of hertz tells exactly how many vibrations are made in one second.

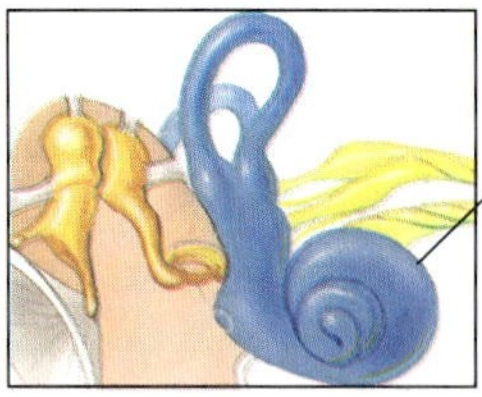

Inner ear: The inner ear includes a spiral-shaped part that turns vibrations into a signal the brain can understand.

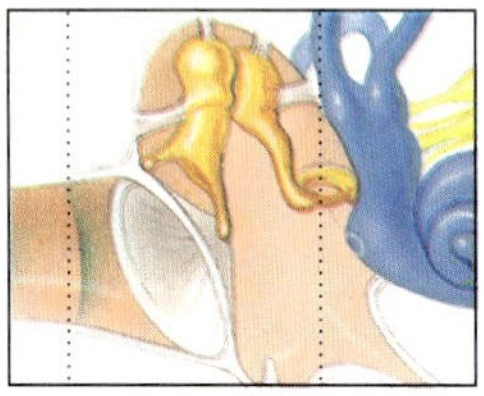

Middle ear: The middle ear includes the eardrum and the three tiny bones next to it. The eardrum and bones send sound vibrations to the inner ear.

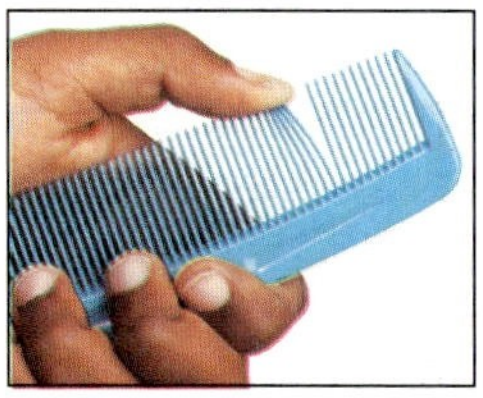

Pitch: Pitch is the highness or lowness of a sound. It's caused by the sound's frequency. The higher the frequency, the higher the pitch.